AF599580

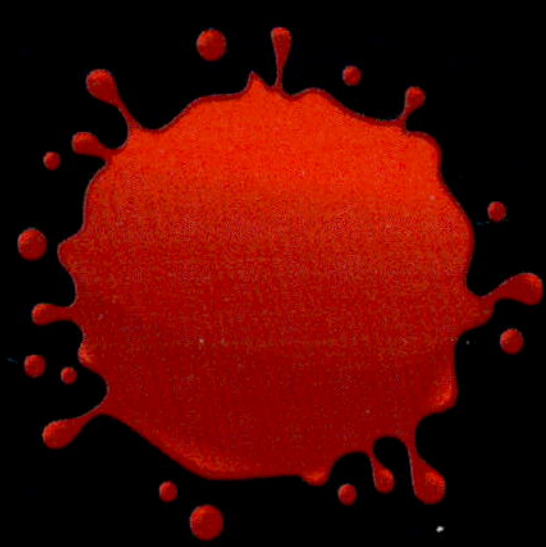

Nicola Lopetz

A Crabtree Seedlings Book

Crabtree Publishing
crabtreebooks.com

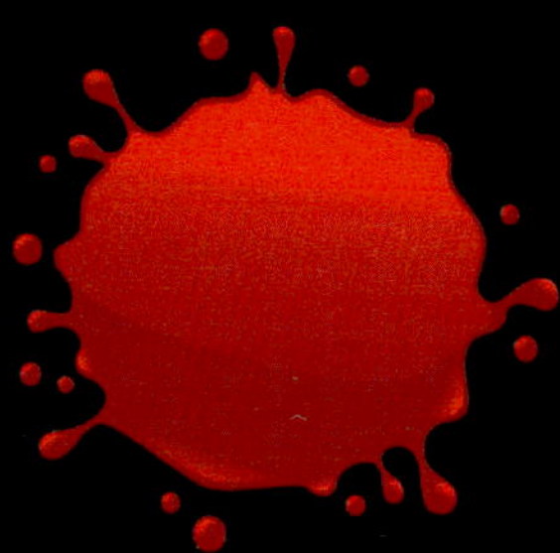
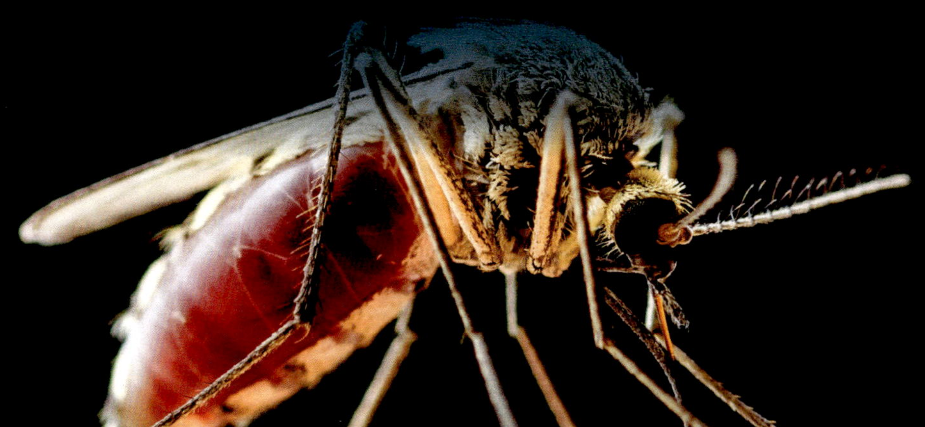

TABLE OF CONTENTS

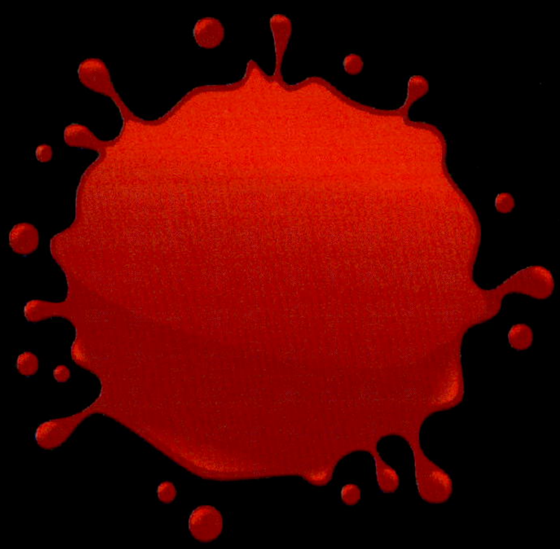

WE WANT TO SUCK YOUR BLOOD!

Animals have many different ways of getting food. **Carnivores** hunt **prey**. Herbivores eat only plants. Some animals survive by drinking blood.

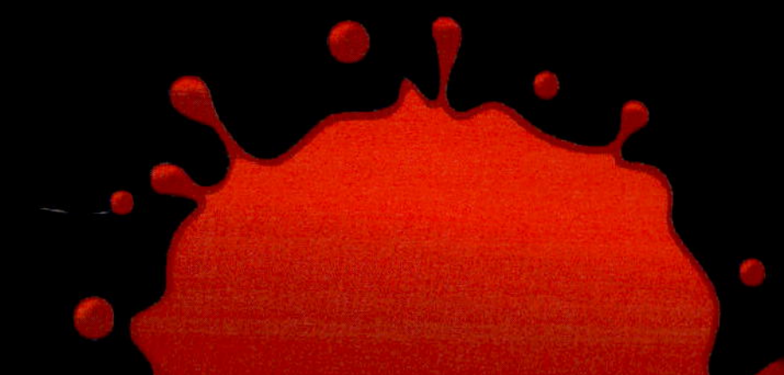

CREEPY OR COOL?

A tick attaches itself to a **host** with a barbed body part near its mouth. Then it buries its head in the host's skin and starts sucking—sometimes for up to two hours!

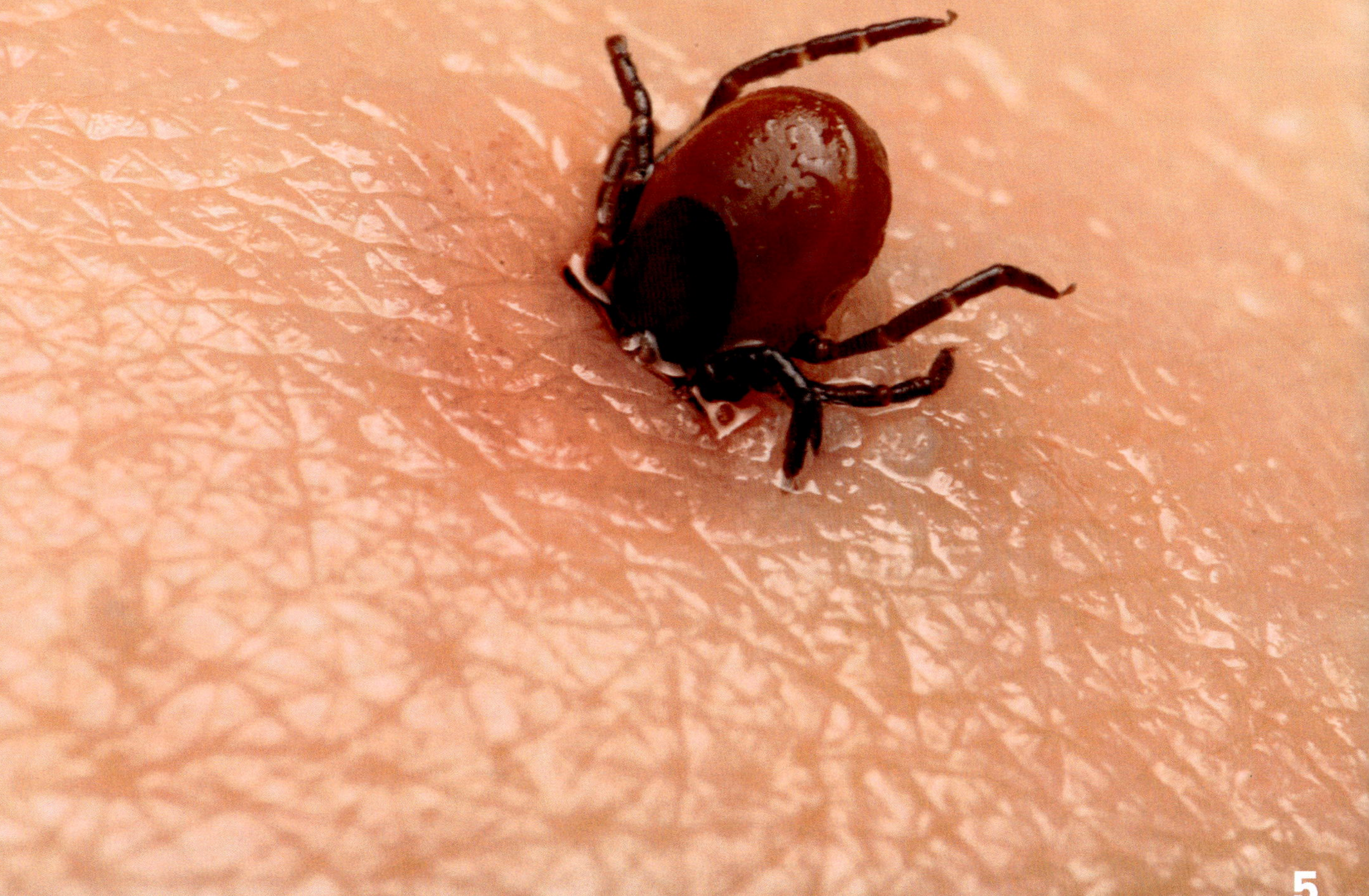

Leeches are a kind of worm. They have a suction cup at each end of their body to hold onto their host. Leeches that drink blood have sharp teeth to saw open their host's skin.

Creepy or Cool?

Leeches have a chemical in their saliva that stops blood from clotting. This has made them useful for doctors to use on some patients.

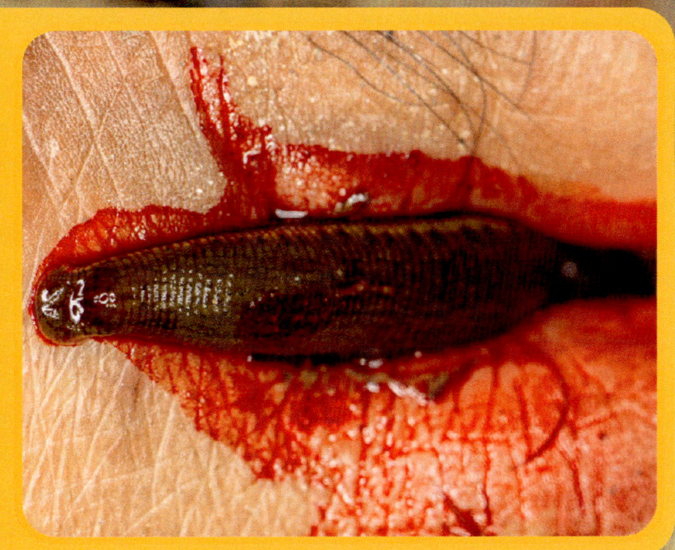

Most leeches live in freshwater, but some can be found on land or in the ocean.

Blood... delicious and nutritious?

Blood is full of **proteins** and **vitamins**. That's good news for animals that feed on only blood.

BLOODTHIRSTY BIRDS

There aren't any birds that dine on only blood, but there are a few that drink blood as part of their diet.

Oxpeckers drink the blood from open wounds on large African mammals. Oxpeckers can be useful, too. They eat ticks, fleas, and maggots from the mammals' hides.

Creepy or Cool?

Oxpeckers also eat the mammals' earwax and dandruff.

Birds do not have teeth. Many birds have sharp beaks for stabbing, cutting, or tearing.

Hood mockingbirds live on the Galápagos Islands. They sometimes feed on the blood of wounded seabirds.

The sharp-beaked ground finch, or vampire finch, lives on the Galápagos Islands too. It uses its beak to slash cuts in seabirds and then slurps the blood.

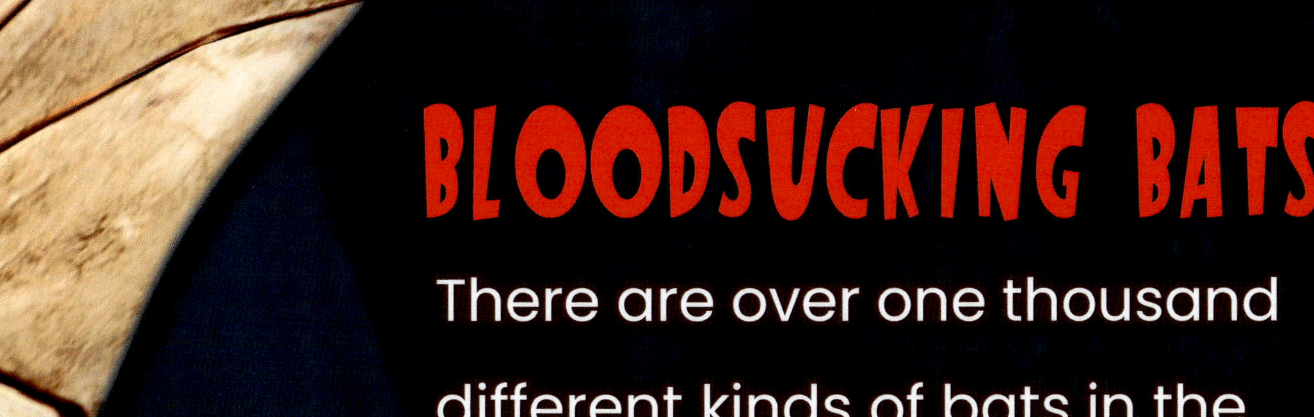

BLOODSUCKING BATS

There are over one thousand different kinds of bats in the world. Only three kinds drink blood. Phew!

CREEPY OR COOL?

The three kinds of bats that drink blood are:

1. The common vampire bat
2. The white-winged vampire bat
3. The hairy-legged vampire bat

Vampire bats live in Mexico, Central America, and South America.

Common vampire bats sneak onto large animals. They nip the skin and lap up the blood.

Like all vampire bats, white-winged vampire bats feed on only blood.

INFECTED INSECTS

Mosquitoes carry a number of serious **diseases**. They spread disease when they transfer **infected** blood to different people.

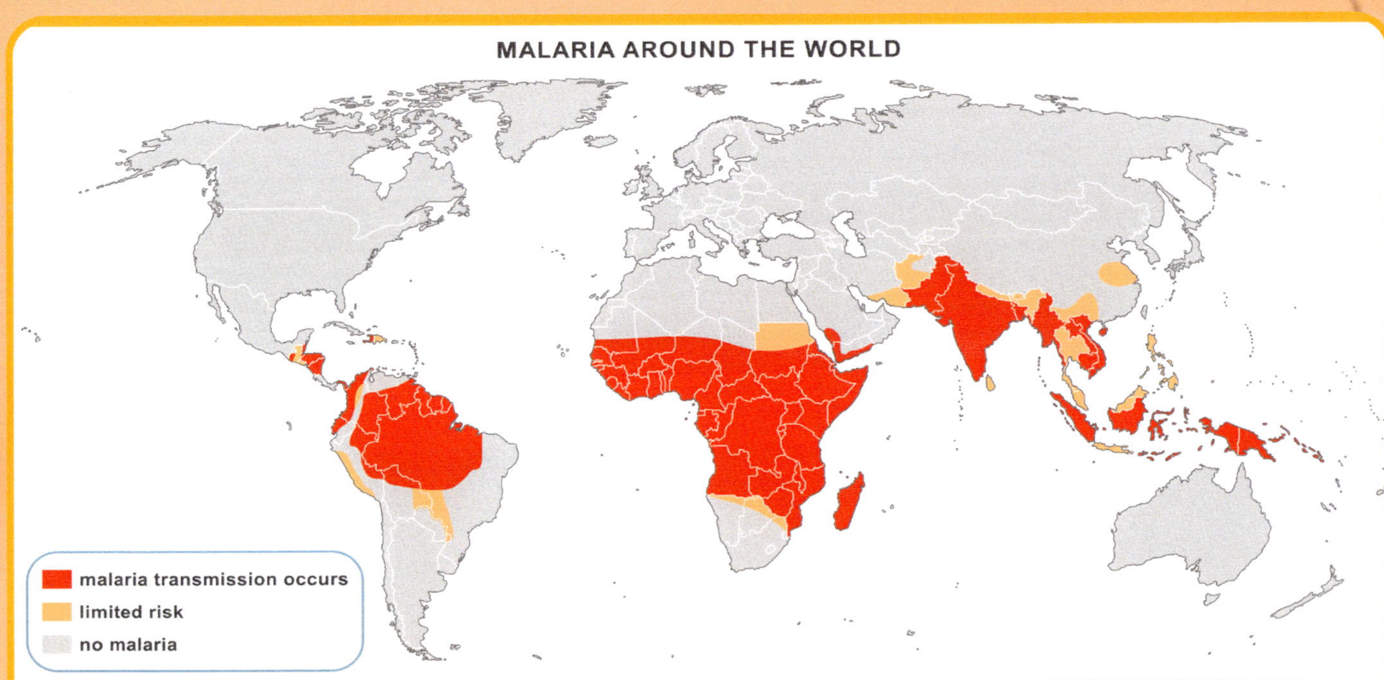

Malaria is one of the deadly diseases spread by mosquitoes.

CREEPY OR COOL?

Only female mosquitoes drink blood. They need it to make eggs.

A mosquito pushes its **proboscis** into the skin of its victim. It sucks blood up through the proboscis, using it like a straw.

Fleas are tiny, but they sure can make life miserable. They feed on warm-blooded animals— including humans!

Fleas have mouthparts for piercing skin and sucking blood.

"Night-night, sleep tight, don't let the bedbugs bite!"

Bedbugs live in pillows and mattresses. They usually feed at night while their victim is sleeping.

Head lice live in human hair and feed on blood from the scalp.

FAMISHED FISH

A sea lamprey looks harmless—until you look into its mouth. YIKES! A sea lamprey's mouth is like a powerful suction cup filled with circular rows of teeth.

CREEPY OR COOL?

Whether you think they are cool or creepy, the creatures in this book drink blood to stay alive.

GLOSSARY

carnivores (KAR-nuh-vorz): Animals that catch and eat other animals

diseases (duh-ZEEZ-iz): Illnesses

host (HOHST): An animal or plant on which another animal lives and feeds

infected (in-FEKT-id): Contains harmful germs

prey (PRAY): Animals hunted and eaten by other animals

proboscis (pruh-BAHS-is): An animal's long snout or feeding tube

proteins (PROH-teenz): Substances in food needed for a healthy body

vitamins (VYE-tuh-minz): Substances in food needed for a healthy body

INDEX

School-to-Home Support for Caregivers and Teachers

This book helps children grow by letting them practice reading. Here are a few guiding questions to help the reader build his or her comprehension skills. Possible answers appear here in red.

Before Reading

- **What do I think this book is about?** *I think this book is about animals that suck blood. I think this book is about vampires.*
- **What do I want to learn about this topic?** *I want to learn what animals are bloodsuckers. I want to learn if there are any real vampires.*

During Reading

- **I wonder why...** *I wonder why doctors sometimes use leeches on some patients. I wonder why mosquitoes drink the blood of humans.*
- **What have I learned so far?** *I have learned that birds do not have teeth. I have learned that there are over 1,000 different kinds of bats in the world and only three kinds drink blood.*

After Reading

- **What details did I learn about this topic?** *I have learned that only female mosquitoes drink blood, they need it to make eggs. I have learned that head lice live in human hair and feed on blood from the scalp.*
- **Read the book again and look for the glossary words.** *I see the word **infected** on page 18, and the word **proboscis** on page 19. The other glossary words are found on page 23.*

Crabtree Publishing

crabtreebooks.com 800-387-7650

Print book version produced jointly with Blue Door Education in 2022

Written by Nicola Lopetz

Hardcover 978-1-4271-6163-5
Paperback 978-1-4271-6175-8

Printed in Canada/052024/CPC20240522

Published in Canada
Crabtree Publishing
616 Welland Avenue
St. Catharines, Ontario
L2M 5V6

Published in the United States
Crabtree Publishing
347 Fifth Avenue
Suite 1402-145
New York, NY 10016

Photo credits:www.shutterstock.com - www.istock.com - www.dreamstime.com -Cover photo © Michael Lynch, title Page: istock.com/nechaev-kon; blood splat throughout ©resnak; page 5 ©Aksenova Natalya; page 6-7 © Stephane Bidouze, page 7 inset ©sydeen; pages 8-9 ©Africa Studio; pages 10-11 ©Stuart G. Porter; pages 12-13 ©Harold Stiver, map ©Serban Bogdan; pages 14-15 ©shutterstock.com/Nathapol Kongseang - Michael Lynch, page 16 map ©Serban Bogdan, page 17 © Gcarter2; pages 18-19 ©mrfiza, map ©Peteri; page 20©Sarah2, page 21 lice inset ©Protasov AN, child ©JPC-PROD; pages 22 ©Drwo_Male, page 22 inset photo ©TiiT Hunt

Library and Archives Canada Cataloguing in Publication
Title: Bloodsuckers / Nicola Lopetz
Names: Lopetz, Nicola, author.
Description: Series statement: Creepy but cool | "A Crabtree seedlings book". | Includes index. | Previously published in electronic format by Blue Door Publishing FL in 2015.
Identifiers: Canadiana (print) 20210200758 | Canadiana (ebook) 20210200766 | ISBN 9781427161635 (hardcover) | ISBN 9781427161758 (softcover) | ISBN 9781427161871 (HTML) | ISBN 9781427161994 (EPUB) | ISBN 9781427162113 (read-along ebook)
Subjects: LCSH: Bloodsucking animals—Juvenile literature. | LCSH: Bloodsucking insects—Juvenile literature.
Classification: LCC QL756.55 .L66 2022 | DDC j591.5/3—dc23

Library of Congress Cataloging-in-Publication Data
Names: Lopetz, Nicola, author.
Title: Bloodsuckers / Nicola Lopetz.
Description: New York : Crabtree Publishing, [2022] | Series: Creepy but cool - a Crabtree seedlings book | Includes index.
Identifiers: LCCN 2021018454 (print) | LCCN 2021018455 (ebook) | ISBN 9781427161635 (hardcover) | ISBN 9781427161758 (paperback) | ISBN 9781427161871 (ebook) | ISBN 9781427161994 (epub) | ISBN 9781427162113
Subjects: LCSH: Bloodsucking animals--Juvenile literature.
Classification: LCC QL756.55 .L67 2022 (print) | LCC QL756.55 (ebook) | DDC 591.5/3--dc23
LC record available at https://lccn.loc.gov/2021018454
LC ebook record available at https://lccn.loc.gov/2021018455